The Open University

Mathematics: A Second Level Course

Linear Mathematics   Unit 15

AFFINE GEOMETRY AND CONVEX CONES

Prepared by the Linear Mathematics Course Team

The Open University Press

The Open University Press Walton Hall Bletchley Bucks

First published 1972
Copyright © 1972 The Open University

Designed by the Media Development Group of the Open University.

Printed in Great Britain by
Martin Cadbury Printing Group

SBN 335 01107 1

This text forms part of the correspondence element of an Open
University Second Level Course. The complete list of units in the
course is given at the end of this text.

For general availability of supporting material referred to in this text,
please write to the Director of Marketing, The Open University, Walton
Hall, Bletchley, Buckinghamshire.

Further information on Open University courses may be obtained from
the Admissions Office, The Open University, P.O. Box 48, Bletchley,
Buckinghamshire.

Contents

Set Books

D. L. Kreider, R. G. Kuller, D. R. Ostberg and F. W. Perkins, *An Introduction to Linear Analysis* (Addison–Wesley, 1966).

E. D. Nering, *Linear Algebra and Matrix Theory* (John Wiley, 1970).

If is essential to have these books; the course is based on them and will not make sense without them.

Conventions

Before working through this correspondence text make sure you have read *A Guide to the Linear Mathematics Course*. Of the typographical conventions given in the Guide the following are the most important.

The set books are referred to as:

> **K** for *An Introduction to Linear Analysis*
> **N** for *Linear Algebra and Matrix Theory*

All starred items in the summaries are examinable.

References to the Open University Mathematics Foundation Course Units (The Open University Press, 1971) take the form *Unit M100 3, Operations and Morphisms*.

15.0 INTRODUCTION

In this unit we shall look at some geometrical aspects of vector spaces. One reason for doing this is that the geometry of real vector spaces differs from school Euclidean geometry. This difference will help to clarify the distinction between the vector spaces which we have been studying so far in the course, and the more highly structured spaces called Euclidean spaces which we shall study in *Unit 16, Euclidean Spaces I*. Another reason is that some of these geometrical properties, particularly those relating to linear inequalities, which were discussed in *Unit M100 6, Inequalities,* are useful in understanding the theory of linear programming, which is the subject of *Unit 18, Linear Programming*. And finally, a study of the geometry associated with vector spaces may help you to gain an intuitive feeling for some of their properties.

The way that different types of geometry arise from different definitions of equivalence was introduced in *Unit M100 35, Topology*. The type of geometry we shall be looking at here is called *affine* geometry. Unlike school Euclidean geometry (which, for 3 dimensions, is a good representation of the space we live in), it lacks the concepts of length and angle. Mathematically, therefore, it is a simpler structure; on the other hand it may be a little difficult at first to get used to doing without these two concepts. In Euclidean geometry, rectangles and squares can be distinguished as special types of parallelograms; but in affine geometry this distinction cannot be made, since we do not have the concept of angle to distinguish the parallelogram from the rectangle, nor the concept of length to distinguish the rectangle from the square. On the other hand we can still distinguish the parallelogram from a general quadrilateral, because affine geometry does have the concept of parallelism.

Euclidean equivalence (congruence)

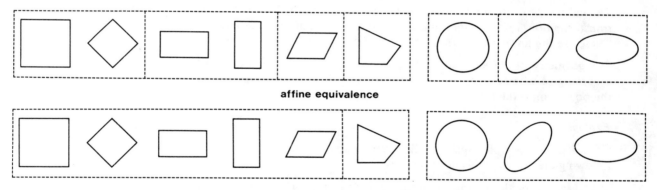

affine equivalence

equivalence classes of figures in Euclidean and affine geometry

15.1   VECTOR GEOMETRY

15.1.0   Introduction

In this section we start by considering a suitable geometry for vector spaces and we extend this to *affine geometry* by stripping the zero vector of its special significance. We achieve this by extending the idea of vector space so that in addition to linear transformations we allow a new type of operation called *translation*. Basically translation is not a new concept. We have met it before in the solution of linear problems. For example, the solution set of $T\xi = \eta$ is $\xi_p + K$ where ξ_p is a particular solution and K is the kernel of T. In the terms of this section we would say that the solution set of $T\xi = \eta$ is the image of the subspace K under translation by ξ_p.

We investigate our new geometry first for two-dimensional vector spaces and then for vector spaces of arbitrary finite dimension. A vector space looked at in this way is called an *affine space*.

15.1.1   The Affine Plane

In *Unit M100 35, Topology*, sub-section 35.2.0, we defined a geometry to be the study of the behaviour of points in R^2 under a group of transformations (i.e. functions $R^2 \longrightarrow R^2$ that are one-to-one and onto) of a given type. We noticed that this set up an equivalence relation defined on the subsets of R^2:

> $A \sim B$ if there is a transformation T of the set such that $T(A) = B (A, B \subset R^2)$.

The set of all transformations of a given type is so chosen that it forms a group with composition of mappings as the binary operation. We use this last property as our jumping-off point.

The natural geometry of a vector space V is the one in which we define the transformations to be automorphisms of V (i.e. isomorphisms from V to V). These are the linear transformations of maximal rank.

The geometry of a vector space V then classifies objects (subsets of V) according to whether there is an automorphism of V which maps one of the objects onto the other.

Example

Consider R^2 as a vector space. In the "vector space geometry" of R^2 the following are typical equivalence classes:

(i)   Lines through the origin (subspaces of R^2 of the form $\{\alpha: \alpha = t\alpha_0\} = \langle \alpha_0 \rangle$).

(ii)   Finite lines with one end point at the origin (i.e. subsets of R^2 of the form $\{\alpha: \alpha = t\alpha_0, t \in [0, 1]\}$).

(iii)   Circles and ellipses with centre at the origin (i.e. subsets of R^2 of the form $\{x\alpha + y\beta: x^2 + y^2 = 1\}$ where $\{\alpha, \beta\}$ is a basis for R^2).

We can verify that each of the above forms an equivalence class as follows.

(i)   Consider two lines through the origin

$$\langle \alpha_0 \rangle \quad \text{and} \quad \langle \beta_0 \rangle.$$

We can choose a vector γ_0 such that $\{\alpha_0, \gamma_0\}$ and $\{\beta_0, \gamma_0\}$ are bases of R^2; then

$$\alpha_0 \longmapsto \beta_0$$
$$\gamma_0 \longmapsto \gamma_0$$

defines an automorphism of R^2 which maps $\langle \alpha_0 \rangle$ onto $\langle \beta_0 \rangle$.

(ii)   Let the two lines be

$$A = \{\alpha \colon \alpha = t\alpha_0, \, t \in [0, 1]\}$$

and

$$B = \{\beta \colon \beta = t\beta_0, \, t \in [0, 1]\}.$$

The automorphism in (i) maps A onto B.

(iii)   Consider the two ellipses

$$E_1 = \{x\alpha_1 + y\beta_1 \colon x^2 + y^2 = 1\}$$
$$E_2 = \{x\alpha_2 + y\beta_2 \colon x^2 + y^2 = 1\}$$

Since $\{\alpha_1, \beta_1\}$ and $\{\alpha_2, \beta_2\}$ are bases we can define an automorphism by

$$\alpha_1 \longmapsto \alpha_2$$
$$\beta_1 \longmapsto \beta_2$$

This maps E_1 onto E_2.

Although this leads to an interesting geometry, it suffers from a serious drawback: the special place that the origin occupies. In other words vector space geometry is restricted by the fact that a vector space has a "tagged" element, the zero vector.

To see this we look at some ways of representing geometrical objects in the vector space R^2. The most natural objects to study in a vector space are the subspaces and the simplest of these are one-dimensional subspaces. In R^2 these are represented by lines through the origin. A line through the origin in the vector space R^2 is thus a set of the form

$$l_0 = \{(x, y) \colon ax + by = 0\} \qquad (a, b \in R)$$

This subspace can be described in two ways. Firstly, l_0 is spanned by the vector $(-b, a)$, i.e.

$$l_0 = \langle(-b, a)\rangle = \{(-tb, ta) \colon t \in R\} = \{(x, y) \colon ax + by = 0\}$$

Alternatively, l_0 is the kernel of the linear functional

$$\phi \colon (x, y) \longmapsto ax + by \qquad (x, y) \in R^2.$$

Thus we have two dual characterizations of a line through the origin:

(i)   the subspace spanned by an element $\xi \in R^2$,

(ii)   the subspace annihilated by an element $\phi \in \widehat{R^2}$.

This duality features prominently in our study of affine spaces and in *Unit 18, Linear Programming*. You should note that just as the choice of basis vector is not unique—any non-zero scalar multiple of it would do—so ϕ is not unique, for the same reason.

Now, the line l_0 is very special because it must contain the zero vector. To avoid any reference to the zero vector we must consider the set of all lines in the plane; they represent subsets of R^2 defined by

$$l = \{(x, y) \colon ax + by = c\} \qquad (a, b, c \in R).$$

In our new geometry we will require that the set of all lines be an equivalence class. Let us first see how we can generalize the description of lines through the origin to arbitrary lines.

The line l can be written in terms of linear functionals as

$$l = \{(x, y) \colon \phi(x, y) = c\} = \phi^{-1}(c).$$

Any two lines of this form with the same ϕ but different values of c are parallel—they have no point in common.

The dual description requires a basis vector; but, in general, l is not a (vector) subspace and therefore has no basis. At this stage we introduce a new type of transformation. The transformation $R^2 \longrightarrow R^2$ defined by

$$(x, y) \longmapsto (x + h, y + k),$$

where $(h, k) \in R^2$, is called a *translation*. We enlarge our group of transformations by allowing translations; since translations are not linear they do not respect the status of the zero vector.

In the case of the line l, we see that it is the image under translation of the line

$$l_0 = \{(x, y): \phi(x, y) = 0\}.$$

For, if $\gamma = (h, k) \in l$, then the translation

$$T_\gamma: \xi \longmapsto \gamma + \xi \qquad (\xi = (x, y) \in R^2)$$

maps l_0 onto l.

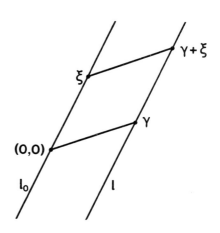

This now gives us the dual description

$$l = \gamma + \langle(-b, a)\rangle$$

since $l_0 = \langle(-b, a)\rangle$.

In full

$$l = \{\xi: \xi = \gamma + \eta, \eta \in \langle(-b, a)\rangle\}$$

i.e.

$$l = \{\xi: \xi = \gamma + t(-b, a), t \in R\}$$

In this last form t is a *parameter* and the whole is called the *parametric representation* of the line l. Thus, we are able to characterize an arbitrary straight line l in R^2 by either of two dual representations:

(i) the image obtained by translating the subspace spanned by an element of R^2,
(ii) the inverse image of an element of R under the linear functional ϕ.

The first yields the parametric representation of l and the second gives a single equation which describes l.

Examples

1. Find the line through the two points $(1, 2)$ and $(-3, 4)$.

The parametric representation is very convenient here. We can take $\gamma = (-3, 4)$, and then the line through $(0, 0)$ parallel to the required line passes through $(1, 2) - \gamma = (4, -2)$. The required line is therefore

$$\{(x, y): (x, y) = (-3, 4) + t(4, -2), t \in R)\}$$

or $\{(-3 + 4t, 4 - 2t): t \in R\}$

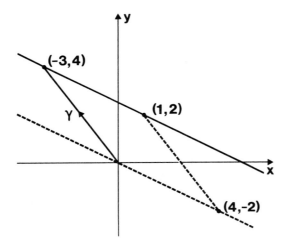

2. Find a line through (1, 2) parallel to the line $x + 2y = 3$.

The given line is $\phi(x, y) = 3$ with $\phi : (x, y) \longmapsto x + 2y$. Any parallel line has the form $\phi(x, y) = c$. Since (1, 2) is on the line, we must have $\phi(1, 2) = c$; so the line is

$$\{(x, y) : x + 2y = 5\}.$$

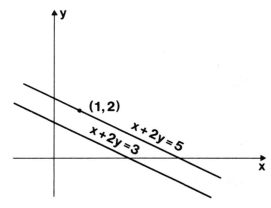

3. Give a parametric representation for the line $x + 2y = 3$.

The parallel line l_0 through the origin is $x + 2y = 0$ and a point on this line is $(-2, 1)$; i.e. $l_0 = \langle(-2, 1)\rangle$. A point on the original line is $\gamma = (1, 1)$. A parametric representation is

$$\{(x, y) : (x, y) = \gamma + t(-2, 1) = (1 - 2t, 1 + t), t \in R\}$$

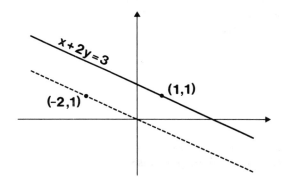

4. A line has the parametric representation

$$\{(-3 + 4t, 4 - 2t) : t \in R\}$$

What is its equation?

The line is $(-3, 4) + \langle(4, -2)\rangle$. A parallel line through the origin is $\langle(4, -2)\rangle$. A linear functional annihilating all points (i.e. vectors) on this line is

$$\phi : (x, y) \longmapsto 2x + 4y$$

The equation of the line is therefore $\phi(x, y) = c$, with c chosen so that $(-3, 4)$ lies on the line: the equation is

$$2x + 4y = 10$$

or

$$x + 2y = 5.$$

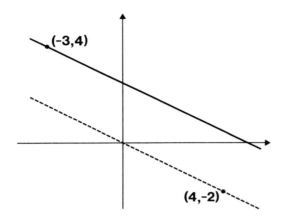

We now define an *affine transformation* of R^2 as a mapping of the form

$$K = L \square T_\alpha$$

defined by

$$\xi \longmapsto L(\xi) + \alpha \qquad (\xi \in R^2)$$

where L is a linear isomorphism of R^2 and T_α is a translation of R^2.

Since the set of all affine transformations forms a group (see Exercise 6, page C11) we define *affine geometry* of R^2 to be the study of subsets of R^2 and their properties which are invariant under affine transformations. In particular, a line l in R^2 is an affine concept, since if $l = \gamma + l_0$ and $K = L \square T_\alpha$

$$\begin{aligned}
K(l) &= K(\gamma + l_0) \\
&= L(\gamma + l_0) + \alpha \\
&= L(\gamma) + L(l_0) + \alpha \\
&= (L(\gamma) + \alpha) + L(l_0)
\end{aligned}$$

which is again a line. When we wish to be specific we refer to a line in this context as an *affine line* and to R^2 as an *affine plane*. This will be generalised to higher dimensions in the next sub-section.

In the affine geometry of R^2 the zero vector has no particular significance and so all lines in R^2 are equivalent. Similarly all ellipses in R^2 are equivalent. This corresponds to our elementary plane geometry where no individual point has any special significance.

We now proceed to consider the simplest affine concept of all. Each element of R^2 is represented by a point in the plane, and it is easily seen that the set of points forms an equivalence class in affine geometry.

Just as each line in R^2 has a dual pair of representations, so each point can be described either by the sum of two vectors or as the intersection of a pair of lines. However, not every pair of lines determines a point—the lines may be parallel. Dual representations may not yet seem very useful or profitable, but when we come to linear inequalities we will find that the dual representations are extremely important.

Example

Find the points of intersection (if any) of the pairs of lines

(i) $x + y = 2$ (i.e. the line $\{(x, y): x + y = 2\}$)

   $x - y = 0$

(ii) $x + y = 2$

      $x + y = 0$

(iii) $x + y = 2$

      $2x + 2y = 4$

In case (i),

      $x = 1$

      $y = 1$

so that the point of intersection is (1, 1).

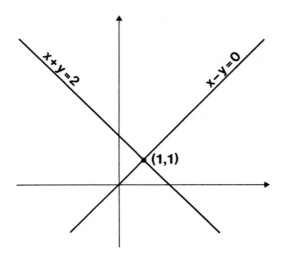

In case (ii), there is no point of intersection because the equations are inconsistent, and the lines are parallel.

In case (iii), the two equations are linearly dependent and there are an infinite number of solutions because the lines coincide.

Exercises

1. For the following pairs of lines, find the point of intersection, if any. If there is no unique point of intersection, draw appropriate conclusions about the lines.

    (i) $x + 2y = 3$ (ii) $x + 2y = 3$ (iii) $x + 2y = 3$

        $2x + 4y = 4$ $2x - y = -4$ $2x + 4y = 6$

2. Find a parametric representation of the line containing the two points (2, 3) and $(-1, -2)$.

3. Find the equation of the line $\{(-1 + 2t, t): t \in R\}$.

4. Find the equation of a line through (3, 4) parallel to the line $x - y = 0$.

5. Give a parametric representation of the line $x + y = 5$.

6. Prove the statement in the text (page C10) that the set of affine transformations forms a group. (Optional)

Solutions

    1.   (i)   No intersection—the lines are parallel.

         (ii)   Point of intersection: $(-1, 2)$

        (iii)   The lines coincide.

            For the method, see the last example.

    2. Two possibilities are

$$\{(2 - 3t, 3 - 5t): t \in R\}$$

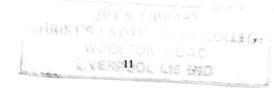

and $\quad \{(-1 + 3t, -2 + 5t) \colon t \in R\}$

For the method, see Example 1.

3. $x - 2y = -1$ or any non-zero multiple of this.
   For the method, see Example 4.

4. $x - y = -1$
   For the method, see Example 2.

5. Two possibilities are

$$\{(5 + t, -t) \colon t \in R\}$$

and $\quad \{(t, 5 - t) \colon t \in R\}$

For the method, see Example 3.

6. Let $K_1 = L_1 \,\square\, T_\alpha$, $K_2 = L_2 \,\square\, T_\beta$, $K_3 = L_3 \,\square\, T_\gamma$ be three affine transformations.

   (a) *Closure* If $\xi \in R^2$, we have

   $$\begin{aligned} K_1 \circ K_2(\xi) &= (L_1 \,\square\, T_\alpha) \circ (L_2 \,\square\, T_\beta)(\xi) \\ &= (L_1 \,\square\, T_\alpha)(L_2(\xi) + \beta) \\ &= L_1(L_2(\xi) + \beta) + \alpha \\ &= L_1 \circ L_2(\xi) + L_1(\beta) + \alpha \\ &= L_1 \circ L_2(\xi) + T_{L_1(\beta) + \alpha} \end{aligned}$$

   (b) *Associativity*

   $$\begin{aligned} (K_1 \circ K_2) \circ K_3 &= [(L_1 \,\square\, T_\alpha) \circ (L_2 \,\square\, T_\beta)] \circ (L_3 \,\square\, T_\gamma) \\ &= [(L_1 \circ L_2) \,\square\, T_{L_1(\beta) + \alpha}] \circ (L_3 \,\square\, T_\gamma) \\ &= (L_1 \circ L_2 \circ L_3) \,\square\, T_{L_1 \circ L_2(\gamma) + L_1(\beta) + \alpha} \end{aligned}$$

   $$\begin{aligned} K_1 \circ (K_2 \circ K_3) &= (L_1 \,\square\, T_\alpha) \circ [(L_2 \,\square\, T_\beta) \circ (L_3 \,\square\, T_\gamma)] \\ &= (L_1 \,\square\, T_\alpha) \circ [(L_2 \circ L_3) \,\square\, T_{L_2(\gamma) + \beta}] \\ &= (L_1 \circ L_2 \circ L_3) \,\square\, T_{L_1(L_2(\gamma) + \beta) + \alpha} \\ &= (L_1 \circ L_2 \circ L_3) \,\square\, T_{L_1 \circ L_2(\gamma) + L_1(\beta) + \alpha} \, . \end{aligned}$$

   (c) *Identity* $I \,\square\, T_0$ acts as the identity (as you can check).

   (d) *Inverse* $L^{-1} \,\square\, T_{-L^{-1}(\alpha)}$ is the inverse of $L \,\square\, T_\alpha$ (as you can check). Remember that L is an isomorphism and hence L^{-1} exists.

15.1.2   Linear Manifolds

We wish to generalize the ideas of sub-section 15.1.1 to spaces of higher dimension. A " three-dimensional " affine space, for example, will contain as proper subsets not only points and lines but also planes. In general, an affine space will contain points, lines, planes and analogous subsets with 3 or more " dimensions ". Such subsets are called *linear manifolds*.

READ from **1 | Vector Geometry** *on page* **N 220** *to the end of that page.*

We consider first the affine geometry of R^3; in the next passage from **N** we will see the generalization to arbitrary vector spaces.

Just as in the two-dimensional case considered earlier, there are two dual descriptions for any linear manifold L. One of them is by means of a system of equations:

$$L = \{\xi \colon \xi \in R^3, \phi_1(\xi) = c_1, \ldots, \phi_r(\xi) = c_r\}$$

For example, in R^3, the two equations

$$\phi(\xi) = 3, \qquad \phi \colon (x, y, z) \longmapsto x + y + z,$$

and $\quad \psi(\xi) = 1, \qquad \psi \colon (x, y, z) \longmapsto 2x + y - 2z,$

separately define planes (as you saw in *Unit M100 15, Differentiation II*), but the set of points that satisfies both equations simultaneously is the intersection of those two planes, which is a line. That is

$$\{(x, y, z): x + y + z = 3\}$$

is a plane, but

$$l = \{(x, y, z): x + y + z = 3 \quad \text{and} \quad 2x + y - 2z = 1\}$$

is a line.

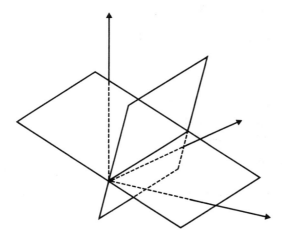

The second description can be obtained from the first by solving the system of equations. In the case of the above system, Hermite normal form brings the system into the form

$$\left.\begin{aligned} x - 3z &= -2 \\ y + 4z &= 5 \end{aligned}\right\} \quad \text{i.e.} \quad \left.\begin{aligned} x &= 3z - 2 \\ y &= -4z + 5 \end{aligned}\right\}$$

so that the solution set is

$$l = \{(3z - 2, -4z + 5, z): z \in R\}$$

or

$$l = (-2, 5, 0) + \langle(3, -4, 1)\rangle.$$

This is our alternative description of the line *l*. Its geometrical interpretation is that the line *l* is obtained from the subspace $\langle(3, -4, 1)\rangle$, which is the line through the origin parallel to *l*, by the translation $(-2, 5, 0)$. The same result can also be interpreted by regarding the original system of equations as a linear problem, $(-2, 5, 0)$ as a particular solution, and $\langle(3, -4, 1)\rangle$ as the solution set of the associated homogeneous problem.

In the above example the space was 3-dimensional, we had 2 equations, and our linear manifold turned out to be one-dimensional. This could have been predicted by using the Dimension Theorem (Theorem 1.6, page **N** 31). Suppose the space has *n* dimensions and there are *s* equations in the system. We may assume its matrix to have rank *s*; for if it did not then either the equations would be inconsistent, in which case the manifold they define is empty, or else some of them are redundant in which case we can delete them. Then the Dimension Theorem tells us that the kernel has dimension $n - s$; the linear manifold defined by the system of equations (being obtained by translating the kernel) therefore also has $n - s$ dimensions.

Sometimes we want to work in the opposite direction from the calculation given above—a linear manifold is described as a translated subspace and we want to describe it in terms of linear equations. In doing this, we use the concept of an annihilator, as we did in sub-section 15.1.1. Suppose,

13

for example, we want a system of equations characterizing the manifold (a line)

$$l = (1, 2, 3) + \langle (4, 5, 6) \rangle.$$

We start by considering the annihilator of the subspace $S = \langle (4, 5, 6) \rangle$. This annihilator, which we denote by $S^\perp$, consists of all linear functionals $[a\ b\ c]$ such that

$$4a + 5b + 6c = 0$$

(See *Unit 12, Linear Functionals and Duality*, sub-section 12.3.3.)

We can pick, say, b and c arbitrarily; then a is determined as $-\frac{5}{4}b - \frac{6}{4}c$. Thus every linear functional in $S^\perp$ has the form

$$[a\ b\ c] = b[-\tfrac{5}{4}\ \ 1\ \ 0] + c[-\tfrac{6}{4}\ \ 0\ \ 1].$$

In other words the linear functionals $[-\frac{5}{4}\ \ 1\ \ 0]$ and $[-\frac{6}{4}\ \ 0\ \ 1]$ form a basis for $S^\perp$. This enables us to specify S completely as the set of all points (vectors) annihilated by the basis functionals, so that we have

$$S = \{(x, y, z): -\tfrac{5}{4}x + y = 0, -\tfrac{6}{4}x + z = 0\},$$

i.e., the subspace S is characterized by the system of equations

$$-\tfrac{5}{4}x + z = 0.$$
$$-\tfrac{6}{4}x + y = 0$$

Finally, to get the corresponding description for l, we translate S by replacing the equations of the form $\phi(\xi) = 0$ by equations of the form $\phi(\xi) = k$ with each k chosen so that the prescribed point $(1, 2, 3)$ is on l, i.e. $k = \phi(1, 2, 3)$. This gives

$$l = \{(x, y, z): -\tfrac{5}{4}x + y = \tfrac{3}{4}, -\tfrac{6}{4}x + z = \tfrac{3}{2}\}$$

as the required description of l by a system of equations. You will find another example of this type of calculation in note (vii) below.

READ all of pages N221 *and* 222.

Notes

(i) *line 3, page* N221 The figure illustrates this.

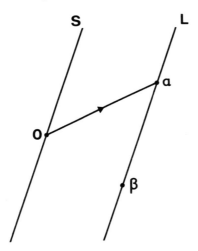

(ii) *line 7, page* N221 " vector in it " means " vector in *L* ".
(iii) *Equation (1.4), page* N221 This is the definition of c_i.

14

(iv)   *line − 12, page* **N221** For example, in the figure, $S_1 \subset S_2$ and so L_1 and L_2 are parallel.

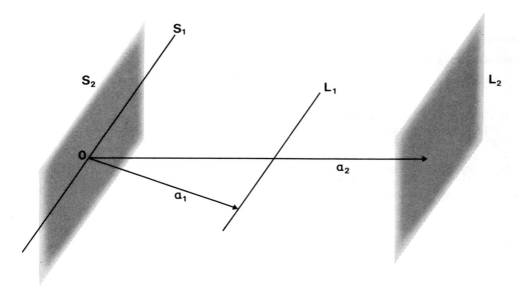

(v)   *line −7, page* **N221**.

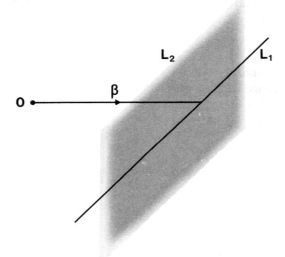

(vi)   *line −19, page* **N222** $S_1 + S_2$ means $\{\alpha_1 + \alpha_2 : \alpha_1 \in S_1, \alpha_2 \in S_2\}$ (see page N21). It is not the same thing as $S_1 \cup S_2$—it is the smallest *subspace* containing both S_1 and S_2.

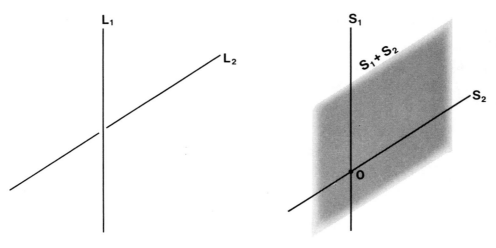

(vii) *line −11, page* **N222** We calculate $(S_1 + S_2)^{\perp}$ by the same method as above. For the linear functional $[a\ b\ c]$ to annihilate the subspace $\langle(2, -1, 1), (1, 0, 2)\rangle$, it must satisfy

$$2a - 1b + 1c = 0.$$
$$1a \qquad + 2c = 0.$$

Reducing to Hermite normal form gives

$$a + 2c = 0$$
$$b + 3c = 0.$$

Thus, $[a\ b\ c] = [-2c\ -3c\ c] = c[-2\ -3\ 1]$, giving $\langle[-2\ -3\ 1]\rangle$, or as **N** prefers $\langle[2\ 3\ -1]\rangle$, as the annihilator.

(viii) *line −9, page* **N222** The point is that the equation of the plane M is $[2\ 3\ -1](x, y, z) = 5$.

(ix) *line −8, page* **N222** Every linear functional that annihilates $S_1 + S_2$ also annihilates the subspace S_1. Hence every member of $(S_1 + S_2)^{\perp}$ is a member of $S_1^{\perp}$, and so $(S_1 + S_2)^{\perp} \subset S_1^{\perp}$. (See also Theorem 4.4, page **N140**.)

Exercises

1. Exercise 1, page **N229**.

2. Exercise 2, page **N229**.

Solutions

1. Answers are given on page **N339**. We indicate the method in more detail for the linear conditions in part (1) *only*.

   The subspace parallel to L_1 is $\langle(1, 1, 1), (2, 1, 0)\rangle$. The annihilator (found by the method illustrated in note (vii) above) is $[1\ -2\ 1]$, i.e. the single linear equation $x_1 - 2x_2 + x_3 = 0$ characterizes the subspace. The equation of the linear manifold is therefore of the form $x_1 - 2x_2 + x_3 = c$, and since $(1, 0, 1)$ is in the manifold we have $1 - 2 \times 0 + 1 = c$, i.e. $c = 2$. Thus the required equation is $x_1 - 2x_2 + x_3 = 2$, or $[1\ -2\ 1](x_1, x_2, x_3) = 2$.

2. Linear conditions for $L_1 \cap L_2$ are immediately found to be

   $$[1\ -2\ 1](x_1, x_2, x_3) = 2$$
   $$[1\ 2\ 2](x_1, x_2, x_3) = 9$$

   These can be written

   $$\begin{bmatrix} 1 & -2 & 1 \\ 1 & 2 & 2 \end{bmatrix} \begin{bmatrix} x_1 \\ x_2 \\ x_3 \end{bmatrix} = \begin{bmatrix} 2 \\ 9 \end{bmatrix}$$

   a solution of which is

   $$x_1 = \tfrac{11}{2} - 6s, \quad x_2 = \tfrac{7}{4} - s, \quad x_3 = 4s.$$

   Therefore

   $$L_1 \cap L_2 = (\tfrac{11}{2}, \tfrac{7}{4}, 0) + s(-6, -1, 4).$$

15.1.3 Affine Closure

Following the method of Example 1 and Exercise 2 of sub-section 15.1.1, we can easily see that any two given points in an affine plane are contained in just one line. In affine spaces of higher dimension this property still holds, but in addition we can also determine other linear manifolds by specifying an appropriate number of points: for example, any three given points, provided that they are not collinear (i.e. do not lie in a 1-dimensional subspace), are contained in just one plane.

On the other hand, if the three points are collinear, then the linear manifold they determine is a line—there is no *unique* plane through them.

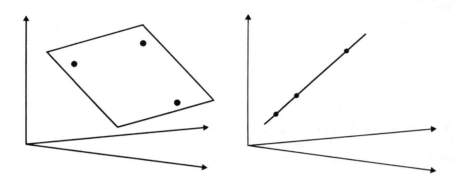

In general, the linear manifold determined in this way by some set of points in an affine space is called the *affine closure* of the set of points.

The simplest case to consider is where one of the points in the set is the origin (the zero vector). Since any linear manifold containing the origin is a subspace, the affine closure of such a set is the smallest *subspace* containing all the points of the set. It is not hard to show that this subspace is just the subspace spanned by the given set of vectors. For example, in R^2, if the given set is

$$\{(1, 1), (0, 0), (-2, -2)\},$$

then the smallest subspace containing it is

$$\langle (1, 1), (0, 0), (-2, -2) \rangle = \langle (1, 1) \rangle.$$

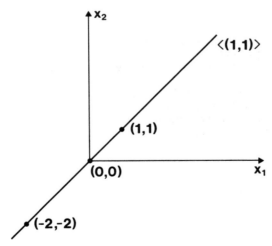

For the general case, the affine closure is not a subspace, but since it is a linear manifold it can be obtained from a subspace by a translation. For example, suppose we want the affine closure of the set $\{(1, 2), (2, 3), (-1, 0)\}$ in R^2.

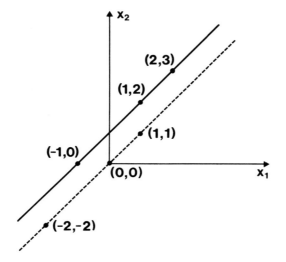

Since $(1, 2)$ is in this manifold, it can be obtained by applying the translation $(1, 2)$ to some subspace S. Then S must contain the points whose images under this translation are the other points in the given set; these points in S are

$$(-1, 0) - (1, 2) = (-2, -2)$$

and

$$(2, 3) - (1, 2) = (1, 1).$$

Thus S is the subspace

$$\langle(0, 0), (-2, -2), (1, 1)\rangle = \langle(1, 1)\rangle$$

and it follows that the affine closure of the original set of points is

$$(1, 2) + \langle(1, 1)\rangle.$$

*READ from page N223 to the end of the statement of **Theorem 1.2** on page N225.*

Notes

(i) *line 9, page* N223 Since $\alpha_i \in L$ and $L = \alpha_0 + S$, we have $\alpha_i - \alpha_0 \in S$; that is, the translation corresponding to the vector $-\alpha_0$ maps L onto S.

(ii) *Equations (1.7) and (1.8), page* N223 These are the key results in this reading passage.

(iii) *line* -13, *page* N223 Notice that to generate a linear manifold of dimension r in an affine space we need at least $r + 1$ points, whereas to generate a subspace of dimension r in a vector space, r vectors are enough. In effect, the $(r + 1)$th vector defining an r-dimensional subspace is the zero vector, which is *ex-officio* a member of every subspace.

(iv) *line* -3, *page* N223 For example, if

$$\alpha_0 = (1, 2), \ \alpha_1 = (2, 3), \ \alpha_2 = (-1, 0)$$

as in our example above, then Equations (1.12) are

$$1c_0 + 2c_1 - 1c_2 = 0$$
$$2c_0 + 3c_1 \qquad = 0$$

giving, with Equation (1.11), a system of equations whose general solution is $(c_0, c_1, c_2) = c(-3, 2, 1)$. Thus, non-trivial linear relations of the form in Equation (1.9) do exist here, and so the dimension of L is less than $r = 2$. In fact, as we showed above, L is a line and its dimension is 1. The linear relations are typified by

$$-3\alpha_0 + 2\alpha_1 + \alpha_2 = 0; \text{ i.e. } \alpha_0 = \tfrac{2}{3}\alpha_1 + \tfrac{1}{3}\alpha_2.$$

(v) *line 2, page* N224 Equations (1.11) and (1.12) form a system with coefficient matrix

$$\begin{bmatrix} 1 & 1 & \cdots & 1 \\ a_{10} & a_{11} & \cdots & a_{1r} \\ \vdots & \vdots & \vdots & \vdots \\ a_{n0} & a_{n1} & \cdots & a_{nr} \end{bmatrix}$$

The homogeneous linear problem has a non-trivial solution if and only if the columns are linearly dependent as vectors in F^{n+1} (F is the field of scalars).

(vi) *line 6, page* N224

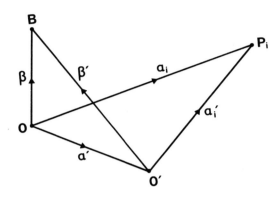

(vii) *Equations (1.7)′ and (1.10)′ page* N224 These equations tell us that Equations (1.7) and (1.10) are invariant under translations. Since they are derived from vector space concepts they are invariant under all affine transformations and therefore define affine concepts. **N** gives them names in the last paragraph on page N224

Note also that line 14 follows using Equation (1.8) and line 18 follows using Equation (1.10),

(viii) *line* −9, *page* N224 For example, if $\alpha_0, \ldots, \alpha_r$ are real (i.e. vectors in **R**) their average is an affine combination. As a further example, if $\alpha_0, \ldots, \alpha_r$ are points in space (representing the affine space R^3) at which objects of masses $m_0, \ldots, m_r$ are placed, their centre of mass $t_0\alpha_0 + \cdots + t_r\alpha_r$ ($t_k = m_k/(m_0 + \cdots + m_r)$) is an affine combination.

(ix) *line 9, page* N225 Do not worry about the meaning of "orthogonal transformations" now. The term will be explained in *Unit 24, Orthogonal and Symmetric Transformations.*

(x) *Theorem 1.1, page* N225 In the statement of the theorem, Λ is an index set which allows us to consider an infinite collection of L_λ. (**N**'s notation is defined on page N6).

The proof requires the result of Theorem 4.1 (page N21) which was omitted from the course; you will not be examined on it.

*READ the statement and proof of **Theorem 1.3** on page* N226.

Exercise

Exercise 3, page N229.

Solution

L will be of the form $(2, 1, 2) + S$, where S is the smallest subspace containing the original point set displaced by $-(2, 1, 2)$; that is $\{(0, 0, 0), (0, 1, -1), (-3, 0, \frac{3}{2})\}$. Thus we have

$$L = (2, 1, 2) + \langle (0, 1, -1), (-3, 0, \tfrac{3}{2}) \rangle.$$

In order that L be parallel to $L_1 \cap L_2$, we require $(-6, -1, 4) \in S$. (see Solution 2 of sub-section 15.1.2).

In fact

$$(-6, -1, 4) = -(0, 1, -1) + 2(-3, 0, \tfrac{3}{2}).$$

15.1.4 Summary of Section 15.1

In this section we defined the terms ·

translation	(page C8)	★ ★
parametric representation	(page C8)	★ ★
affine transformation	(page C10)	★ ★
affine geometry	(page C10)	★
affine line	(page C10)	★
affine plane	(page C10)	★
linear manifold	(page N220)	★ ★ ★
hyperplane	(page N220)	★ ★ ★
affine dependence	(page N224)	★
affine closure	(page N225)	★

Theorems

1. (*1.1*, page N225)
Let $\{L_\lambda : \lambda \in \Lambda\}$ be any collection of linear manifolds in V. Either their intersection is empty or it is a linear manifold.

2. (*1.2*, page N225)
Let S be any subset of V. Let $\bar{S}$ be the set of all affine combinations of finite subsets of S. Then $\bar{S}$ is a linear manifold.

3. (*1.3*, page N226)
The affine closure of S is the set of all affine combinations of finite subsets of S.

Notation

$$T_\gamma \qquad\qquad\qquad \text{(page C8)}$$

15.2   CONVEX SETS

15.2.0   Introduction

In Section 15.1 we saw that a linear manifold can be identified with the solution set of a system of linear equations. In the ensuing work our concern will be systems of *linear inequalities*. For example, the solution set of the system of inequalities in R^2:

$$x_1 \geqslant 0$$
$$x_2 \geqslant 0$$
$$x_1 + x_2 \leqslant 1$$

is bounded by a triangle.

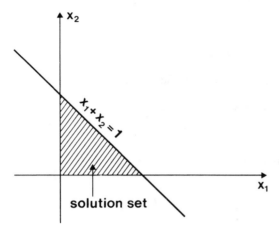

One reason for studying these solution sets is that they are important in linear programming and operational research.

The solution sets of systems of linear inequalities are examples of *convex sets*. Geometrically, they are characterized by the fact that if P_1 and P_2 are any two points of the set, then all the points on the line segment *between* P_1 and P_2 also belong to the set.

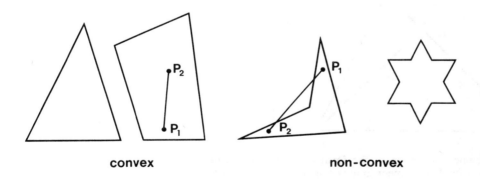

In this section we shall use the ideas of affine geometry to study convex sets in real vector spaces. We shall see that there are many analogies between the properties of convex sets and those of linear manifolds. For our purposes one of the most important is that a convex set, like a linear manifold, can be described in two dual ways—either by a system of conditions (equations or inequalities) specifying the set, or by giving a general formula for the points in it*.

---

* The analogy is only valid for convex sets with linear boundaries such as convex polygons and polyhedra.

21

15.2.1 Convex Linear Combinations

The main new idea that we need in order to define convexity is that of *betweenness*; that is, we need to be able to order the points on a line.

READ from page N227 line 6 to the end of page N228.

Notes

(i) *lines 12-16, page* N227

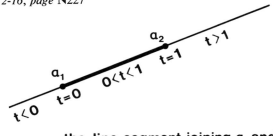

the line segment joining a_1 and a_2

The thick part is the line segment joining α_1 and α_2. $t_1 = 1 - t$ and $t_2 = t$; so $0 \leqslant t \leqslant 1$ is equivalent to $0 \leqslant t_1$ and $0 \leqslant t_2$.
Before continuing we assert that betweenness is an affine concept. Let $L \,\square\, T_\gamma$ be an affine transformation; we write

$$\beta' \text{ for } (L \,\square\, T_\gamma)\beta, \text{ and } \alpha_i' \text{ for } (L \,\square\, T_\gamma)\alpha_i$$

We will demonstrate that Equation (1.14) is invariant under $L \,\square\, T_\gamma$.

$$
\begin{aligned}
(L \,\square\, T_\gamma)\beta &= L(\beta) + \gamma \\
&= (1 - t)L(\alpha_1) + tL(\alpha_2) + \gamma \\
&= (1 - t)\{L(\alpha_1) + \gamma\} + t\{L(\alpha_2) + \gamma\} \\
&= (1 - t)(L \,\square\, T_\gamma)(\alpha_1) + t(L \,\square\, T_\gamma)(\alpha_2)
\end{aligned}
$$

i.e. $\beta' = (1 - t)\alpha_1' + t\alpha_2'$

(ii) *lines* -15 *and* -14*, page* N227 For " $\cap_{\lambda \in A} C_\lambda$ " read "the intersection".
(iii) *Equations (1.15) and (1.16), page* N227 Notice how these equations generalize the definition of a line segment. As an example, suppose $r = 3$ and $\alpha_1, \alpha_2, \alpha_3$ are at the vertices of a triangle. We show that any point inside the triangle is a convex linear combination of $\{\alpha_1, \alpha_2, \alpha_3\}$.

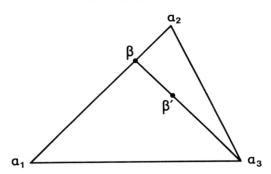

Any point β on the line segment α_1, α_2 is of the form $(1 - t)\alpha_1 + t\alpha_2$ with $t \in [0, 1]$, and any point β' on the line segment joining β and α_3 is of the form

$$(1 - t')\beta + t'\alpha_3$$

where $t' \in [0, 1]$.

Substituting for β, this form becomes

$$(1 - t')(1 - t)\alpha_1 + (1 - t')t\alpha_2 + t'\alpha_3,$$

which is a convex linear combination of α_1, α_2 and α_3, since all 3 coefficients are positive and $(1 - t')(1 - t) + (1 - t')t + t' = 1$.
(iv) *line* -2*, page* N227 Note that in general the corners (if any) of a convex set are called its *vertices*.

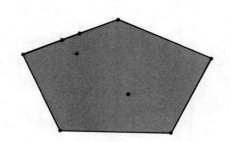

(v) *lines* −6 *to* −1, *page* **N227**
For example, all convex linear combinations of the set of 9 points shown lie in the shaded area or on its boundary. If, however, S is an infinite subset of the plane, we do not always obtain a polygon; e.g. let S be the union of two lines.
(vi) *Proof of Theorem 1.8, page* **N228** The first 3 lines of the proof prove the "if"; the rest proves the "only if", i.e. that if C is convex then every convex linear combination of members of C is itself a member of C. This second part is an example of proof by induction in which the hypothesis is that $P_1, P_2, \ldots, P_{r-1}$ are all true. In Equation (1.17), notice that

$$\sum_{i=1}^{r-1} \frac{t_i}{1-t_r} t_r = t_r, \quad \text{since} \quad \sum_{i=1}^{r-1} t_i = 1 - t_r, \text{ by Equation (1.16).}$$

(vii) *line 17, page* **N228** For example, the shaded region (with its boundary) shown in note (v) is the convex hull of the set of 9 points shown.

Exercises

1. Exercise 4, page N229.

2. Prove that the convex hull of a set of points $\{\alpha_1, \ldots, \alpha_r\}$ is a subset of their affine closure.

3. Prove that the solution set of the system of linear inequalities

    $$ax + by \geqslant c, dx + ey \geqslant f, gx + hy \geqslant i$$

    is a convex set.

4. Exercise 11, page N141. (*Hint* By a theorem of analysis, if f is a continuous real function with domain $[a, b]$ and k is any number between $f(a)$ and $f(b)$ then there is a $c \in [a, b]$ such that $f(c) = k$.)

Solutions

1. $(0, 0)$ is in the convex hull, which is a triangle with vertices at the points of S. The best proof is to solve for t_1, t_2, t_3 satisfying

    $$(0, 0) = t_1(1, 1) + t_2(-6, 7) + t_3(5, -6)$$

    and

    $$t_1 + t_2 + t_3 = 1.$$

    The resulting values for t_1, t_2, t_3, given on page N339, are all greater than zero; therefore $(0, 0)$ is in the convex hull.

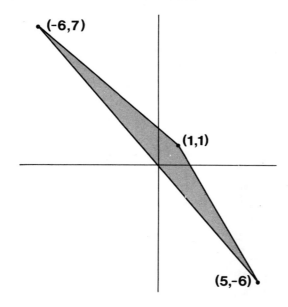

2.   By Equations (1.15), (1.16), (1.7) and (1.8), every convex linear combination of a set of points is an affine combination; the required result then follows by Theorems 1.3 and 1.9.

3.   Let (x_1, y_1) and (x_2, y_2) be any two members of the solution set. Then, if

$$(x_3, y_3) = (1 - t)(x_1, y_1) + t(x_2, y_2)$$

with $t \in [0, 1]$, we have

$$ax_3 + by_3 = (1 - t)(ax_1 + by_1) + t(ax_2 + by_2)$$
$$\geq (1 - t)c + tc = c$$

since $t \in [0, 1]$.

A similar result holds for the other inequalities in the system. Thus (x_3, y_3) is also in the solution set, which is therefore convex.

4.   See page N333. (Note that the hyperplane in this exercise was chosen to be a vector subspace.)

15.2.2   Convex Cones

When we come to apply vector space methods to convex sets, there are some convex sets that have particularly convenient properties, namely *convex cones*. They differ from the examples of convex sets we used in the last sub-section in that they are unbounded sets, i.e. they "extend to infinity" in some directions. In addition they involve reinstating the origin and so we consider convex cones as subsets of real *vector* spaces.

Let us first consider the various types of convex cones in the vector space R^2.

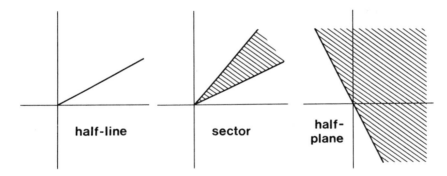

half-line   sector   half-plane

There are 3 distinct types, as illustrated in the diagram. Each satisfies the definition of a convex set, but they also have one further geometrical property: if P is any point in the cone, then every point on the half-line from the origin through P is also in the cone.

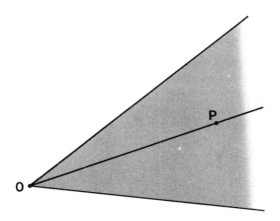

In vector space language, if α is any vector in the cone then every vector $t\alpha$, with $t \geqslant 0$, is also in the cone. That is, these convex cones are closed under multiplication by a non-negative scalar.

This property is not restricted to convex cones in R^2: we may define a convex cone in any real vector space as a convex set that is closed under multiplication by non-negative scalars. One way of constructing such cones in spaces of higher dimension is to take any convex set and to construct all the half lines from the origin through it. Some examples in 3 dimensions are illustrated below.

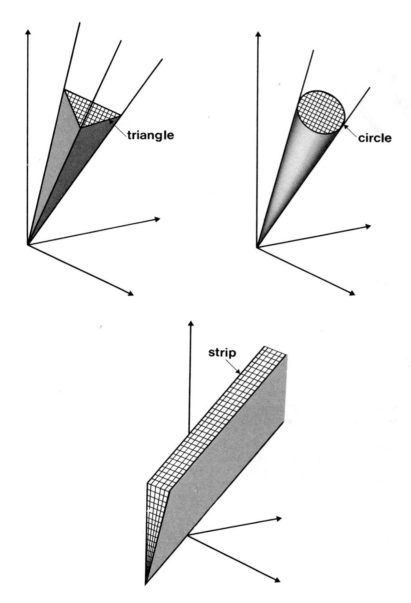

Other types of 3-dimensional cone include the 2-dimensional types we have seen already, and also a half-space (the region on one side of a plane through the origin).

READ from **2 | Finite Cones and Linear Inequalities** *on page N229 to line* -6, *page* N230.

Notes

(i)   *line 7, page* N230 This is equivalent to our definition; for if a set is convex and also closed under multiplication by non-negative scalars, then for any two vectors α_1 and α_2 in the set, the convex linear combination $\frac{1}{2}\alpha_1 + \frac{1}{2}\alpha_2$ is also in the set, and so is the scalar multiple of this $2(\frac{1}{2}\alpha_1 + \frac{1}{2}\alpha_2) = \alpha_1 + \alpha_2$. On the other hand, if the set is closed under multiplication by a non-negative scalar and also under addition, then for any two vectors α_1 and α_2 in it and any number t in [0, 1], the vectors $(1 - t)\alpha_1$, $t\alpha_2$, and hence their sum $(1 - t)\alpha_1 + t\alpha_2$ are also in the set, and the definition of convex set (page N227) is satisfied. Note also that not every cone is convex:

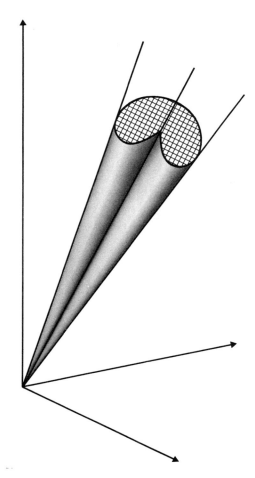

(ii) *line 12, page* N230 "*finite cone*" really means "finitely generated". You might prefer to call it a *pyramid*. The cross-section of a finite cone with n generators is an n-sided polygon.
(iii) *line —7, page* N230 In the set of three cones on page C24, the two on the left are pointed and the one on the right is wedge shaped. In the set shown on page C25, the lower one is wedge shaped; the other two are pointed.

Exercises

1. (i) Which of the following sets are convex cones in R^3?
    (ii) Which are pointed?
    (iii) Which are wedge shaped?
    (iv) Specify a set of generators for each *finite* cone.
        (a) $\{(x_1, x_2, x_3): x_1 \geqslant 0, x_2 \geqslant 0, x_3 \geqslant 0\}$
        (b) $\{(x_1, x_2, x_3): x_1 \geqslant 0, x_2 \geqslant 0\}$
        (c) $\{(x_1, x_2, x_3): x_1 \geqslant 0\}$
        (d) R^3
        (e) $\{(x_1, x_2, x_3): x_1^2 + x_2^2 \leqslant x_3^2; x_3 \geqslant 0\}$

2. Show that the solution set of the system of homogeneous linear inequalities

    $$ax_1 + bx_2 + cx_3 \geqslant 0$$
    $$dx_1 + ex_2 + fx_3 \geqslant 0$$
    $$gx_1 + hx_2 + ix_3 \geqslant 0$$

    in R^3, is a convex cone.

Solutions

1.   (i)   All are convex cones.

    (ii)   (a) and (e) are pointed.

    (iii)   (b) is wedge-shaped.

    (iv)   (a) to (d) inclusive are finite cones: here are some possible sets of generators.

        (a) $\{\alpha_1, \alpha_2, \alpha_3\}$ where $\alpha_1, \alpha_2, \alpha_3$ are the standard basis vectors $(1, 0, 0)$, $(0, 1, 0)$, $(0, 0, 1)$.

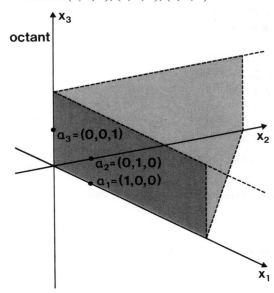

        (b) $\{\alpha_1, \alpha_2, \alpha_3, -\alpha_3\}$. We include $-\alpha_3$, because x_3 can take either sign, and only non-negative multiples of the generators are allowed.

        (c) $\{\alpha_1, \alpha_2, -\alpha_2, \alpha_3, -\alpha_3\}$

        (d) $\{\alpha_1, -\alpha_1, \alpha_2, -\alpha_2, \alpha_3, -\alpha_3\}$

Note that (e) is a circular cone and is not a finite cone.

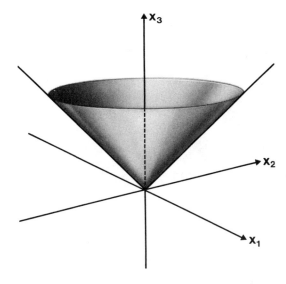

2.   The solution set of the system of inequalities is closed under multiplication by non-negative scalars, and also under addition, and is therefore a convex cone. If (x_1, x_2, x_3) satisfies all the inequalities and k is non-negative, then (kx_1, kx_2, kx_3) also satisfies them all; and if (x_1', x_2', x_3') also satisfies all the inequalities, then so does $(x_1 + x_1', x_2 + x_2', x_3 + x_3')$. In fact we know convexity from sub-section 15.2.1 (Exercise 3).

15.2.3   Summary of Section 15.2

In this section we defined the terms

between	(page N227)	★ ★
convex set	(page N227)	★ ★ ★
convex linear combination	(page N227)	★ ★
vertex	(page C22)	★ ★ ★
convex hull	(page N228)	★ ★ ★
cone	(page N230)	★ ★ ★
convex cone	(page N230)	★ ★ ★
half-line	(page N230)	★ ★
generators	(page N230)	★ ★
finite cone	(page N230)	★ ★ ★
pointed	(page N230)	★

Theorems

1.   (*1.7*, page N227)
The intersection of any number of convex sets is convex.   ★

2.   (*1.8*, page N228)
A set C is convex if and only if every convex linear combination of vectors   ★
in C is in C.

3.   (*1.9*, page N228)
The convex hull of a set S is the set of all convex linear combinations of   ★
vectors in S.

15.3   LINEAR INEQUALITIES

15.3.0   Introduction

In the preceding section and its exercises, we saw two ways of specifying finite cones: (i) a set of generators, (ii) a system of homogeneous linear inequalities whose solution set is the cone. These are analogous to the two dual specifications for a subspace, either by giving a basis, or by giving a system of homogeneous linear equations whose solution set is the subspace. This is only to be expected, since a subspace is a special case of a convex cone. In the case of a subspace S, the equations specifying it have the form

$$\phi_i(\xi) = 0 \qquad (i = 1, 2, \ldots, r)$$

where $\phi_1, \phi_2, \ldots, \phi_r$ are linear functionals forming a basis for the annihilator $S^\perp$ of S. This suggests looking for an analogue for cones of the annihilator for subspaces. This analogue is also a set of linear functionals (i.e. a subset of the dual space), but now we are interested, not in equations of the form $\phi_i(\xi) = 0$, but in inequalities of the form $\phi_i(\xi) \geqslant 0$. Thus we can specify the cone by the set of all linear functionals which map every vector in the cone to a non-negative number. One consequence of this description is the *separating hyperplane theorem* which is proved easily using the "geometry" of the situation, and then tells us a lot about the existence of solutions of a set of linear inequalities.

15.3.1   The Dual Cone

We start by defining the dual cone. This has its application to linear programming where there is a dual problem corresponding to any given set of linear inequalities.

READ from line −5 on page N230, *to line 2 on page* N231, *and from line −14 on page* N231 *to line −6 on page* N231.

Notes

(i)   *line −4, page* N230 $\phi\alpha$ means $\phi(\alpha)$: see page N134.

(ii)   *line −1 page* N230 Notice the use of the isomorphism between $\hat{\hat{V}}$ and V (*Unit 12, sub-section 12.2.3*). This allows us to speak of "vectors (in V) which have non-negative values" instead of "linear functionals on $\hat{V}$ which have ...". In symbols, this definition is: $W^+ = \{\alpha: \phi\alpha \geqslant 0$ for all $\phi \in W\}$.

(iii)   *line −13, page* N231 "*generated by the finite set* ...". For example if C is the cone in $\hat{R^3}$ generated by the linear functionals [1 0 0], [0 1 0] and [1 1 −1], then C^+ is the solution set (in R^3) of the corresponding system of linear inequalities: $x_1 \geqslant 0$, $x_2 \geqslant 0$, $x_1 + x_2 - x_3 \geqslant 0$.

(iv)   *lines −8 to −6, page* N231

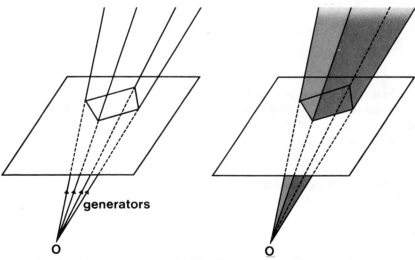

generators

O   O

In the left-hand figure a finite cone C in R^3 and its intersection with a plane are shown. The half-lines contain its generators. In the right-hand figure the shaded faces are planes $\{\alpha: \phi_i\alpha = 0\}$. The polyhedral cone is the intersection of 4 half-spaces: $C = \bigcap_{i=1}^{4} \{\alpha: \phi_i(\alpha) \geqslant 0\} \subset R^3$.

Example

Find a set of generators for the dual cone C^+, where C is the cone in R^3 generated by

$$\{(1, 1, -1), (0, 1, 0), (1, 0, 0)\}.$$

C^+ is the set of all linear functionals ϕ such that $\phi(1, 1, -1)$, $\phi(0, 1, 0)$, and $\phi(1, 0, 0)$ are all non-negative.

Let $\alpha_1 = (1, 1, -1)$, $\alpha_2 = (0, 1, 0)$, $\alpha_3 = (1, 0, 0)$.

A set of generators for C^+ can be found by considering the linear functionals associated with the faces of C, as suggested in the last paragraph of the reading passage.* C has three faces which lie in the planes

$$\langle \alpha_2, \alpha_3 \rangle, \langle \alpha_3, \alpha_1 \rangle, \langle \alpha_1, \alpha_2 \rangle.$$

These have the equations

$$x_3 = 0, \ x_2 + x_3 = 0, \ x_1 + x_3 = 0$$

which correspond to the respective annihilators of the three faces:

$$\langle [0 \quad 0 \quad 1] \rangle, \langle [0 \quad 1 \quad 1] \rangle, \langle [1 \quad 0 \quad 1] \rangle.$$

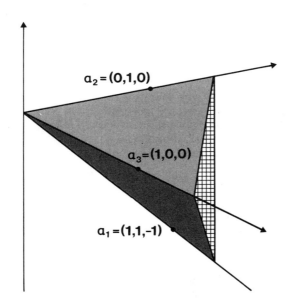

The cone is the intersection of three half spaces, each bounded by one of the planes. The half-space bounded by the plane $x_3 = 0$ is either $x_3 \geqslant 0$ or $x_3 \leqslant 0$; and since it has to include the point $\alpha_1 = (1, 1, -1)$ it must be $x_3 \leqslant 0$. This can be written $\{\alpha: \phi_1 \alpha \geqslant 0 \text{ and } \alpha \in R^3\}$ where $\phi_1 = [0 \quad 0 \quad -1]$, and we take this linear functional as one of our generators for C^+.

In a similar way we find that the other two half-spaces are $x_2 + x_3 \geqslant 0$, and $x_1 + x_3 \geqslant 0$, and corresponding linear functionals are

$$\phi_2 = [0 \quad 1 \quad 1] \quad \text{and} \quad \phi_3 = [1 \quad 0 \quad 1].$$

---

* In general, given a finite cone C with n faces, its dual is a polyhedral cone whose n generators correspond to the n faces of C.

The linear functionals ϕ_1, ϕ_2, ϕ_3 are certainly elements of C^+. To show that they also *generate* C^+ we note that they form a basis for $\widehat{R^3}$, so that *any* linear functional on R^3 has the form $\phi = a_1\phi_1 + a_2\phi_2 + a_3\phi_3$. If a_1, a_2, a_3, are all non-negative, then $\phi(x_1, x_2, x_3)$ is non-negative for all (x_1, x_2, x_3) in C (because ϕ_1, ϕ_2, ϕ_3 have this property), and so ϕ is in C^+. If, however, any of a_1, a_2, a_3 is negative then we can find $\alpha \in C$ such that $\phi\alpha < 0$; for example, if a_1 is negative, we take $\alpha = \alpha_1$ (the first of the generators of C); this gives a negative value for ϕ_1 since $\phi_2\alpha_1 = \phi_3\alpha_1 = 0$ and $\phi_1\alpha_1$ is positive making $(a_1\phi_1 + a_2\phi_2 + a_3\phi_3)(\alpha_1)$ negative. Thus ϕ is not in C^+ if $a_1 < 0$. Similarly, we can show that it is not in C^+ if $a_2 < 0$ or $a_3 < 0$. Thus ϕ is in C^+ if and only if a_1, a_2, a_3 are all non-negative and we conclude that $\{\phi_1, \phi_2, \phi_3\}$ is a set of generators for C^+.

Algebraically, we have shown that the cone, C, determines a system of inequalities

$$x_3 \leqslant 0$$
$$x_2 + x_3 \geqslant 0$$
$$x_1 + x_3 \geqslant 0$$

such that every other linear inequality satisfied by each element of C is a non-negative linear combination of these.

Geometrically, we have specified the cone as the intersection of 3 half-spaces (the first of which is $\{(x_1, x_2, x_3): x_3 \leqslant 0\}$) instead of giving its edges, the generators.

Dual to this, the three vectors $\alpha_1, \alpha_2, \alpha_3$ characterize C^+ as a polyhedral cone; the three linear functionals ϕ_1, ϕ_2, ϕ_3 generate C^+ as a finite cone.

Exercises

1. Exercsie 1, page N237.

2. Exercise 2, page N237.

Solutions

    1. See page N340.

    2. The method is the same as in the worked example. The face contained in the subspace $\langle(1, 0, -1), (0, -1, 1)\rangle$ has the equation $x_1 + x_2 + x_3 = 0$, and the corresponding half-space which includes the cone is $x_1 + x_2 + x_3 \geqslant 0$ (not $\leqslant 0$, since $(1, 1, 0)$ is in the cone). The face contained in the subspace $\langle(1, 1, 0), (0, -1, 1)\rangle$ corresponds to the half-space

$$x_1 - x_2 - x_3 \geqslant 0,$$

and the face contained in the subspace $\langle(1, 1, 0), (1, 0, -1)\rangle$ corresponds to the half-space

$$x_1 - x_2 + x_3 \geqslant 0.$$

The corresponding linear functionals, which form a set of generators for C_1^+, are

$$[1 \quad 1 \quad 1], [1 \quad -1 \quad -1], [1 \quad -1 \quad 1].$$

15.3.2 The Duality of Finite Cones

The results of the previous sub-section indicate that any finite cone C can be described in either of two ways: by giving a set of generators for C or for its dual—the polyhedral cone C^+. Geometrically, the generators of C describe the edges of C and the generators of C^+ describe the faces of C.

READ the statements of Theorems 2.6 and 2.7 on pages N232 and N233.

Notes

(i) *Theorem 2.6, page* **N232** Geometrically, if a cone is the convex hull of a finite set of half-lines (its edges), then it is bounded by a finite set of faces; algebraically, if it has a finite set of generators, then it is the solution set of a finite system of homogeneous linear inequalities. This theorem is illustrated by the Example in sub-section 15.3.1

(ii) *Theorem 2.7, page* **N233** Geometrically, if the cone is bounded by a finite set of faces each of which is a part of a hyperplane (i.e. it is the intersection of a finite set of half-spaces), then it is the convex hull of a finite set of half-lines; algebraically, the solution set of any finite system of homogeneous linear inequalities is the cone generated by some finite set of " basic solutions " of these inequalities (the basic solutions are the generators of the cone).

Example

Find the solution set of the following system of inequalities

$$x_1 + x_3 \geqslant 0$$
$$x_2 + x_3 \geqslant 0$$
$$x_3 \leqslant 0$$

These inequalities constitute a cone $D \subset \widehat{R^3}$ with generators $[1 \quad 0 \quad 1]$, $[0 \quad 1 \quad 1]$, $[0 \quad 0 \quad -1]$. The solution set is the dual cone $D^+ \subset R^3$ whose faces lie in the planes $x_1 + x_3 = 0$, $x_2 + x_3 = 0$, $x_3 = 0$. The lines of intersection of these planes, obtained by solving these equations in pairs, are

$$\langle (1, 0, 0) \rangle \quad \text{for} \quad x_2 + x_3 = 0, x_3 = 0;$$
$$\langle (0, 1, 0) \rangle \quad \text{for} \quad x_1 + x_3 = 0, x_3 = 0;$$
$$\langle (1, 1, -1) \rangle \quad \text{for} \quad x_1 + x_3 = 0, x_2 + x_3 = 0.$$

By *Theorem 2.7*, the solution set, being a polyhedral cone, is a finite cone, and we look for a set of generators lying in these three lines. A generator lying in the first of these lines is either $(1, 0, 0)$ or $(-1, 0, 0)$; the inequality $x_1 + x_3 \geqslant 0$ which holds throughout the cone tells us that $(1, 0, 0)$ is correct. Similarly a generator lying in the second line is $(0, 1, 0)$, not $(0, -1, 0)$ since $x_2 + x_3 \geqslant 0$; and one in the third line is $(1, 1, -1)$, since $x_3 \leqslant 0$. Thus the solution set of the given system of inequalities is the finite cone generated by $\{(1, 0, 0), (0, 1, 0), (1, 1, -1)\}$: in other words

$$(x_1, x_2, x_3) = t_1(1, 0, 0) + t_2(0, 1, 0) + t_3(1, 1, -1)$$
$$(t_1 \geqslant 0, t_2 \geqslant 0, t_3 \geqslant 0).$$

This is just the cone C we started with in the example in sub-section 15.3.1

Exercise

The last example can be formulated as:
Find a set of generators for the solution set of

$$\phi_1(\xi) \geqslant 0$$
$$\phi_2(\xi) \geqslant 0$$
$$\phi_3(\xi) \geqslant 0$$

where $\xi = (x_1, x_2, x_3)$ and $\phi_1 = [1 \quad 0 \quad 1]$, $\phi_2 = [0 \quad 1 \quad 1]$, $\phi_3 = [0 \quad 0 \quad -1]$. The set of linear functionals $\{\phi_1, \phi_2, \phi_3\}$ is a basis in $\widehat{R^3}$. Find its dual basis in R^3.

Compare this dual basis with the generators for the solution set calculated in the above example. (*Hint* Recall from *Unit 12* that dual bases are related by $\phi_i(\alpha_j) = \delta_{ij}$, i.e. the matrix whose rows represent ϕ_1, ϕ_2, ϕ_3 is the inverse of the matrix whose columns represent $\alpha_1, \alpha_2, \alpha_3$.)

Solution

The basis in $\widehat{R^3}$ is $\{[1 \quad 0 \quad 1], [0 \quad 1 \quad 1], [0 \quad 0 \quad -1]\}$ and it is dual to the basis $\{(1, 0, 0), (0, 1, 0), (1, 1, -1)\}$ in R^3, which is identical with the set of generators for the solution set.

15.3.3 The Separating Hyperplane Theorem

We now come to a result which is used in the theory of linear programming. It is discussed in some detail in **N**, but we are concerned with only the general idea.

The result which we shall study is the *separating hyperplane theorem*. The geometrical content of the theorem is this: if *B* is any point outside a convex set, then there exists a hyperplane with *B* on one side of it and the entire convex set on the other. The theorem is intuitively obvious in R^2, where the hyperplane is a line; and in R^3, where the hyperplane is just a plane.

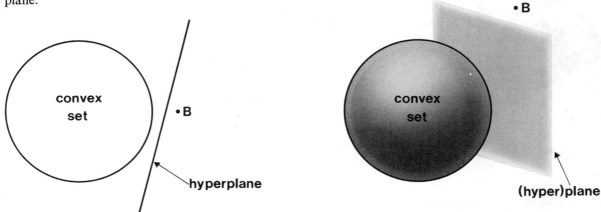

READ the statement of Theorem 2.9 on page N234. Its proof is optional.

Notes

(i) *Theorem 2.9, page N234* The *dual* transformation $\hat{\sigma}: \hat{V} \longrightarrow \hat{U}$ is defined by $\hat{\sigma}: \psi \longmapsto \psi \circ \sigma$ (see page N142). A simpler version may be stated as follows:

> "If *C* is a finite cone in a vector space *V* and β is a vector not in *C*, then there exists a linear functional ψ on *V* such that $\psi(\alpha) \geqslant 0$ for all α in *C*, and $\psi(\beta) < 0$. That is, the hyperplane $\{\eta: \psi(\eta) = 0\}$ divides *V* into two parts, one containing β and the other containing *C*. If $\beta \in C$, then no such ψ exists."
>
> *Proof* Since the cone *C* is finite, it is polyhedral, by Theorem 2.6. That is, there is a finite cone *K* in $\hat{V}$ such that
>
> $$C = K^+ = \{\alpha: \tilde{\alpha}(\phi) = \phi(\alpha) \geqslant 0, \phi \in K\}$$
>
> Since β is not in *C*, we therefore cannot have $\phi(\beta) \geqslant 0$ for all $\phi \in K$; there is at least one ψ in *K* such that $\psi(\beta) < 0$. This is the functional whose existence the theorem asserts.

Let *C* be any finite cone in a vector space *V*, with generators $\{\gamma_1, \ldots, \gamma_n\}$. Then we can regard the formula

$$(t_1, \ldots, t_n) \longmapsto \eta = t_1\gamma_1 + \cdots + t_n\gamma_n$$

for a general element of *C* as a linear transformation from R^n to *V*, which maps the cone generated by the standard basis in R^n to the cone *C*. **N** generalizes this idea slightly by taking an arbitrary *n*-dimensional vector space *U* in place of R^n, and an arbitrary basis $\{\alpha_1, \ldots, \alpha_n\}$ in it. The linear transformation that maps this basis in *U* to the generators $\{\gamma_1, \ldots, \gamma_n\}$ (which need not be a basis) in *V* is denoted by σ, and it maps the cone *P* generated by $\{\alpha_1, \ldots, \alpha_n\}$ to the cone *C* generated by $\{\gamma_1, \ldots, \gamma_n\}$.

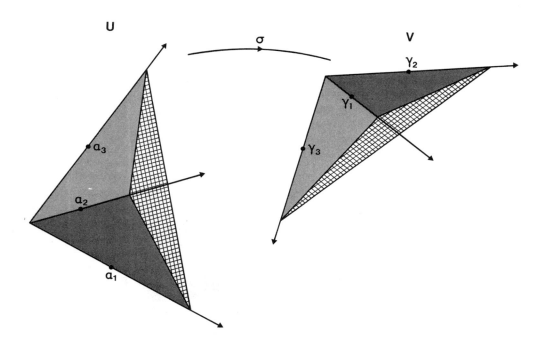

Then, of the two alternatives stated by **N**, (1) states that $\beta \in C$ and (2) (which occurs when $\beta \notin C$) states that there is a linear functional ψ on **V** such that $\psi(\beta) < 0$ and $\hat{\sigma}(\psi) \in P^+$. Now $\hat{\sigma}(\psi) : \xi \longrightarrow \psi(\sigma(\xi))$ is a linear functional on $\hat{U}$ so that $\hat{\sigma}(\psi) \in P^+$ means $\psi(\eta) \geqslant 0$ for all η in **C**. The hyperplane $\{\eta : \psi(\eta) = 0\}$ in **V** "separates" β from **C**.

(ii) *line 14, page* N234 (optional) "there is a $\psi \in \hat{V}, \ldots$" Since $\sigma(P)$ is a polyhedral cone, there is a finite cone $D \subset \hat{V}$ such that $\sigma(P) = D^+$, i.e. $\sigma(P) = \{\eta : \psi(\eta) \geqslant 0, \psi \in D\}$. $\beta \notin \sigma(P)$ implies the existence of a particular $\psi \in D \subset \hat{V}$ such that $\psi\beta < 0$, whereas $\psi\sigma(P) \geqslant 0$.

Exercises

1. Exercise 9, page N238.
    (*Hint* Is (4, 3) in the cone generated by (2, 1), (5, -5) and (-7, -6)? A non-negative solution is one with $x_1 \geqslant 0$, $x_2 \geqslant 0$ and $x_3 \geqslant 0$.)

2. Exercise 7, page N238. (The definition of C_2 is given in Exercise 3 on page N238. $X \geqslant 0$ means that all the entries in the matrix X are greater than or equal to zero.)

Solutions

1. The question asks us, in effect, to show that (4, 3) is not a positive linear combination of (2, 1), (5, -5) and (-7, -6); i.e., that it does not lie in the cone they generated.

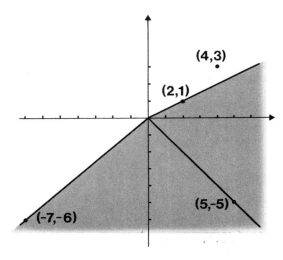

Theorem 2.9 tells us that we can do this by finding a linear functional that is negative for (4, 3) but positive or zero for (2, 1), (5, −5) and (−7, −6). The diagram suggests using a linear functional that is zero at (2, 1) e.g. [1 −2] or [−1 2], and by trying [1 −2] we see that it has the required property. Another way of putting this is that by subtracting twice the second equation from the first we get $15x_2 + 5x_3 = -2$, which obviously has no non-negative solution. Other linear functionals will do too, for example [2 −3].

2. This is similar to Exercise 1. We shall show that (1, 1, 0) is not a positive linear combination of the columns of A; that is, it is not in the cone C_2 they generate. Theorem 2.9 suggests looking for a linear functional $Y = [y_1 \quad y_2 \quad y_3]$, giving a negative image for B and non-negative images for the columns of A. To make the image of B negative we want $y_1 + y_2 < 0$; to make the images of the columns of A positive it is sufficient to take y_3 large, say 100, and $y_1 - y_2 > 0$. A suitable linear functional is thus $Y = [0 \quad -1 \quad 100]$. By Theorem 2.9 there can be no X such that $X \geqslant 0$ and $AX = B$. (With any X satisfying $AX = B$ we have $YAX = YB$, i.e., $[99 \quad 1 \quad 99]X = -1$, so X cannot be greater than or equal to 0. Another way to do it, of course, is to solve the equations, obtaining the solution

$$X = \begin{bmatrix} -2 \\ -1 \\ 2 \end{bmatrix}, \text{ where } X \not\geqslant 0.)$$

15.3.4 Summary of Section 15.3

In this section we defined the terms

dual cone   (page **N230**) ★ ★ ★

polyhedral cone   (page **N231**) ★ ★ ★

dual transformation   (page **N142**) ★ ★ ★

Theorems

1. (**2.6**, page **N232**) ★ ★ ★
Every finite cone is polyhedral.

2. (**2.7**, page **N233**) ★ ★ ★
A polyhedral cone is finite.

3. (**2.9**, page **N234**) ★ ★ ★
Let $A = \{\alpha_1, \ldots, \alpha_n\}$ be a basis of the vector space U and let P be the finite cone generated by A. Let σ be a linear transformation of U into V and let β be a given vector in V. Then one and only one of the following two alternatives holds: either
(1) there is a $\xi \in P$ such that $\sigma(\xi) = \beta$, or
(2) there is a $\psi \in \hat{V}$ such that $\hat{\sigma}(\psi) \in P^+$ and $\psi\beta < 0$.

4. (page **C33**) ★ ★ ★
If C is a finite cone in a vector space V and β is a vector not in C, then there exists a linear functional ψ on V such that $\psi(\alpha) \geqslant 0$ for all α in C, and $\psi(\beta) < 0$.

Notation

$X \geqslant 0$ (page **C34**)

15.4 SUMMARY OF THE UNIT

The main aim of this unit has been to discuss the concepts of affine geometry and convexity.

The first section removes the special significance of the origin from a vector space, looks at the implications of this and investigates the structure of the resulting space, the affine space. We introduced the concept of a linear manifold which can represent the solution set of a system of linear equations, but in the second section we looked at the solution sets of systems of linear inequalities, which are examples of convex cones. The third section deals with dual cones, culminating in the separating hyperplane theorem which is the subject of the last sub-section of the unit.

Definitions

translation	(page **C**8)	★	★	
parametric representation	(page **C**8)	★	★	
affine transformation	(page **C**10)	★	★	
affine geometry	(page **C**10)	★		
affine line	(page **C**10)	★		
affine plane	(page **C**10)	★		
linear manifold	(page **N**220)	★	★	★
hyperplane	(page **N**220)	★	★	★
affine dependence	(page **N**224)	★		
affine closure	(page **N**225)	★		
between	(page **N**227)	★	★	
convex set	(page **N**227)	★	★	★
convex linear combination	(page **N**227)	★	★	
vertex	(page **C**22)	★	★	★
convex hull	(page **N**228)	★	★	★
cone	(page **N**230)	★	★	★
convex cone	(page **N**230)	★	★	★
half-line	(page **N**230)	★	★	
generators	(page **N**230)	★	★	
finite cone	(page **N**230)	★	★	★
pointed	(page **N**230)	★		
dual cone	(page **N**230)	★	★	★
polyhedral cone	(page **N**231)	★	★	★
dual transformation	(page **N**142)	★	★	★

Theorems

Only three-star theorems are listed.

1. (*2.6*, page **N**232)
Every finite cone is polyhedral. ★ ★ ★

2. (*2.7*, page **N**233)
A polyhedral cone is finite. ★ ★ ★

3. (*2.9*, page **N**234)
Let $A = \{\alpha_1, \ldots, \alpha_n\}$ be a basis of the vector space U and let P be the finite cone generated by A. Let σ be a linear transformation of U into V and let β be a given vector in V. Then one and only one of the following two alternatives holds: either   ★ ★ ★
(1) there is a $\xi \in P$ such that $\sigma(\xi) = \beta$, or
(2) there is a $\psi \in \hat{V}$ such that $\hat{\sigma}(\psi) \in P^+$ and $\psi\beta < 0$.

4. (page **C**33)
If C is a finite cone in a vector space V and β is a vector not in C, then there exists a linear functional ψ on V such that $\psi(\alpha) \geqslant 0$ for all α in C, and $\psi(\beta) < 0$.   ★ ★ ★

Notation

T_γ	(page **C**8)
$X \geqslant 0$	(page **C**34)

15.5   SELF-ASSESSMENT

Self-assessment Test

This Self-assessment Test is designed to help you test your understanding of the unit. It can also be used, together with the summary of the unit, for revision. The answers to these questions will be found on the next non-facing page. We suggest that you complete the whole test before looking at the answers.

1. Describe briefly (at most 50 of *your own* words) the main differences between affine geometry and Euclidean geometry.

2. Given a line $2x + 3y = 5$ in R^2:
    (i)   write down the equation of a parallel line through $(4, -2)$;
    (ii) write down a parametric representation of the original line.

3. What is the equation of the plane
    $(0, 0, 1) + \langle(0, 1, 0), (1, 0, 1)\rangle$ in R^3?

4. What is the equation of the affine closure of the set $\{(0, 1), (1, 0)\}$ in R^2?

5. Is the origin a member of the convex hull of the set $\{(1, 0), (0, 1), (1, -1)\}$ in R^2?

6. Is the set $\{(x, y, z): x + y + z = 0\}$ a convex cone in R^3?

7. Find a set of generators for the solution set of

    $$x_1 + x_2 \geqslant 0$$
    $$x_1 - x_2 \leqslant 0$$

    in R^2.

8. Draw a diagram illustrating the separating hyperplane theorem for convex cones in R^2.

9. Formulate and prove a result generalizing the result of the exercise of sub-section 15.3.2 to the case where $\{\phi_1, \phi_2, \phi_3\}$ is any basis in $\hat{R}^3$. Does this result generalize to $\hat{R}^n$?

Solutions to Self-assessment Test

1. Affine geometry studies the way subsets of a vector space behave under affine transformations without distinguishing any connection between vectors other than linear dependence. Euclidean geometry allows the definition of angle between vectors and the comparison of lengths of linearly independent vectors.

2. (i) The required equation has the form $2x + 3y = k$.

$$2 \times 4 + 3 \times (-2) = 2$$

   i.e. $k = 2$, and so the required line has equation $2x + 3y = 2$.
   (ii) One point in the line is $(1, 1)$, so the line is

$$(1, 1) + \langle(-3, 2)\rangle \qquad \text{or} \qquad \{(1 - 3t, 1 + 2t) : t \in R\}$$

3. Let the annihilator of $\langle(0, 1, 0), (1, 0, 1)\rangle$ be $\langle[a \quad b \quad c]\rangle$.
   Then

$$b = 0$$
$$a + c = 0$$

   and so the annihilator is generated by $[-1 \quad 0 \quad 1]$. The plane $\langle(0, 1, 0), (1, 0, 1)\rangle$ therefore has the equation

$$z = x.$$

   The plane we are looking for has the equation

$$x - z = k$$

   where $k = 0 - 1 = -1$
   $\therefore \qquad x - z = -1$

4. The affine closure has the form $(0, 1) + S$ where S is the smallest subspace containing the original "point set" displaced by $-(0, 1)$, i.e. $\{(0, 0), (1, -1)\}$.
   So we have

$$(0, 1) + \langle(1, -1)\rangle.$$

5. Suppose $(0, 0) = t_1(1, 0) + t_2(0, 1) + t_3(1, -1)$ and $t_1 + t_2 + t_3 = 1$. Then $(0, 0)$ is in the convex hull if and only if the above equations yield a solution such that $t_1, t_2, t_3 \geqslant 0$.
   Now

$$t_1 + t_3 = 0$$
$$t_2 - t_3 = 0$$

   implies $t_1 = -t_3$, so we must require $t_1 = t_3 = 0$. This implies $t_2 = 0$ and so we cannot satisfy all the conditions, i.e. $(0, 0)$ is not in the convex hull.

6. The set $\{(x, y, z): x + y + z = 0\}$ is a plane in R^3 and thus a subspace. It is therefore a fortiori a convex cone.

7. Let

$$\alpha = [1 \quad 1],$$
$$\beta = [-1 \quad 1].$$

   $\{\alpha, \beta\}$ is a basis for $\widehat{R^2}$.

Then the generators of the solution set $\{\xi: \xi \in R^2, \alpha\xi \geqslant 0, \beta\xi \geqslant 0\}$ are the elements of the dual basis in R^2, i.e. $\{\xi, \eta\}$ where

$$\alpha\xi = 1 \qquad \beta\xi = 0$$
$$\alpha\eta = 0 \qquad \beta\eta = 1$$

$\therefore$ $\left.\begin{array}{l} \xi = (\tfrac{1}{2}, \tfrac{1}{2}) \\ \eta = (-\tfrac{1}{2}, \tfrac{1}{2}) \end{array}\right\}$ is a set of generators.

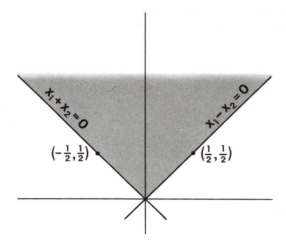

8.

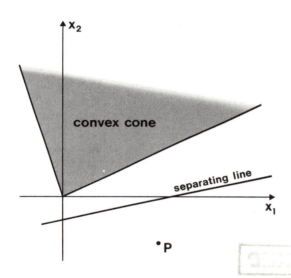

9. The general result for R^3 is:

If $\{\phi_1, \phi_2, \phi_3\}$ is a basis for $\widehat{R^3}$ and $\{\alpha_1, \alpha_2, \alpha_3\}$ is the dual basis in R^3, then the solution set C of the system of inequalities $\phi_1(\xi) \geqslant 0, \phi_2(\xi) \geqslant 0, \phi_3(\xi) \geqslant 0$, is the cone generated by $\{\alpha_1, \alpha_2, \alpha_3\}$. In other words, the cones generated by $\{\phi_1, \phi_2, \phi_3\}$ and $\{\alpha_1, \alpha_2, \alpha_3\}$ are dual to each other.

Proof One edge of the cone C is in $\{\xi: \phi_1(\xi) = \phi_2(\xi) = 0\}$ which is the line $\langle \alpha_3 \rangle$, by the formulas $\phi_i(\alpha_j) = \delta_{ij}$. The generator lying in this line might be either α_3 or $-\alpha_3$ (or both) but the condition

$$\phi_3 \text{ (generator)} \geqslant 0$$

requires us to pick α_3. In a similar way we can show that α_1 and α_2 are generators.

The result does generalize: if $\{\phi_1, \ldots, \phi_n\}$ is a basis for $\widehat{R^n}$ and $\{\alpha_1, \ldots, \alpha_n\}$ is its dual basis in R^n, then the cones generated by these two sets of vectors are dual to each other. The proof follows the same lines as above. (Note that $n-1$ faces meet at each edge.)

LINEAR MATHEMATICS